Post Harvest Processing Technology

NIPA® GENX ELECTRONIC RESOURCES & SOLUTIONS P. LTD.
New Delhi-110 034

About the Author

Dr. Suresh Chandra is working as Professor in the Department of Processing and Food Engineering, College of Technology, Sardar Vallabhbhai Patel University of Agricultural and Technology, Meerut (UP). He obtained his B.Tech. in Ag. Engg. from GBPUAT, Pantnagar; M.Tech. in Post Harvest Engineering and Technology from AMU Aligarh, and Ph.D. in Agric. Process and Food Engineering from SVPUAT, Meerut. He has more than 20 year's service experience in teaching, research and extension. Dr. Chandra published more than 100 research paper, 12 books, 8 practical manuals, 3 training bulletins and various book chapters. He has been conferred with various prestigious awards in his carrier of teaching. He is reviewers of various national and international journals. He has guided 10 M.Tech. and 05 Ph.D. scholars and many more in his guidance. He established Agro-Processing Centre and Food Processing Unit in the University. He has handled many projects funded by ICAR/RKVY/UPCAR and Deptt. of Agriculture, U.P as a PI and Co-PI. He has organized various seminar/symposia/webinar and entrepreneurship development training for farmers, women and rural unemployed persons, and delivered invited talks as resource person and guest lecturers as subject expert in external institutions. He has participated in various summer/winter schools and short courses organized by ICAR-Institutions/SAUs.

Post Harvest Processing Technology

The Protocol Series, Volume 08

Suresh Chandra

Professor

Department of Processing and Food Engineering

College of Technology Sardar Vallabhbhai Patel

University of Agriculture and Technology

Meerut-250110 (UP) India

NIPA® GENX ELECTRONIC RESOURCES & SOLUTIONS P. LTD.

New Delhi-110 034

NIPA® GENX ELECTRONIC
RESOURCES & SOLUTIONS P. LTD.

101,103, Vikas Surya Plaza, CU Block
L.S.C. Market, Pitam Pura, New Delhi-110 034
Ph : +91-11-43860225, Mob.: +91 9717133558, 9540816132
E-mail: newindiapublishingagency@gmail.com
Website: www.nipaersources.com

Print ISBN: 978-93-58877-82-3
ebook ISBN: 978-93-58877-86-1

Composed and Designed by NIPA®.

Foreword

The Indian Council of Agricultural Research (ICAR) has played the vital role in transforming the agricultural higher education to produce highly skilled human resource equipped with advance knowledge and capable of fulfilling the agricultural needs of the country. National Education Policy (NEP-2020) of India represents a comprehensive and ambitious vision for the future of education in the country. There is an urgent need to revise and align the agricultural education system in the country in line with NEP-2020 norms to build a competent human resource for undertaking education, research and extension activities at diversified, ecologically sustainable and economically viable agricultural and allied sectors with integration of technology and innovation.

Enhancing quality of human resource is a pre-requisite for implementing and upgrading research programs, developing technologies, evolving institutional arrangements to face challenges and to harness opportunities in the global perspective. Moreover, in the agriculture and allied sectors, such aspects are still more important due to the rapid changes occurring in challenges as well as in technologies. There is an urgent need of adequately trained manpower, that is professionally competent, socially sensitive and ethically strong to provide the technical backstopping in the changing resource constraints scenario in terms of land and water availability, and sustain agricultural growth with challenges of climatic adversities.

Skill enhancement courses are essential requirements for any programme (UG-Certificate/ UG Diploma or Degree). These will be choice based; student can choose the areas of skill enhancement from a bouquet of skill enhancement modules offered by the parent institution. The institutions will develop capabilities for offering such courses. An institution is at liberty to (and in fact, it should) work in partnership with capable organizations/ companies/ NGOs/ progressive entrepreneurs/ farmers for running various skill enhancement programs. The University/ HAEIs may also formulate and offer courses in any other areas as identified by it, based on institutional expertise/ capabilities/ resources. In addition, the skill enhancement courses suggested by the UGC, may also be offered.

I am happy that Dr. Suresh Chandra have taken initiative step to prepare a Skill Enhancement Course-IV on "Post Harvest Processing Technology" shall be beneficial to the Experiential learning students of B.Sc. (Hons) Ag., B.Tech. (Food Technology) & B.Tech. (Agricultural Engineering) and related disciplines which to train the skill development in the field of Post Harvest Processing Technology under Skill Enhancement Course. The manual covers all the practical aspects related to Food Processing. Authors have organized the content in such a way that facilitates the students and teachers both to understand the objectives of the experiments and activities to initiates during the Skill Enhancement Course.

I am confident that the manual will be useful to the students, teachers and all such professionals who are eager to know about the practical approach in the subject.

I congratulate the author for their sincere efforts.

-Sd-

B.R. Singh

Dean, College of Technology

SVPUAT, Meerut (UP)

Preface

The scope of the Skill Enhancement Course would include hands on training in pre-selected vocations. It would involve creation of facilities in the form of instructional farms, model plants or engineering workshops. The training programme would lean heavily on practice in technologically advanced methodologies for production of agricultural crops, fruits, vegetables, ornamental, medicinal, aromatic, forest plants including fiber and fuel crops etc. It would also engage students in integrated farming systems consisting of production of fish, poultry, livestock for milk, eggs, meat and wool, and processing for value added products etc. The other avenues of practical training would include internships in private industrial houses, technology transfer and rural development programmes. The practical exposure would also involve students in preparation of project feasibility and implementation reports, proper methods and procedures for maintenance of records including inventory of materials, maintenance of accounts, and management of the enterprise. The Post Harvest Processing Technology in the area of fruits and vegetables has to match with the production of these perishable commodities to reach the consumers. For a successful food-processing sector, various aspects such as quality control, quality system and quality assurance, and the constituents of the total quality management should function in a horizontal fashion. Processed Foods have to be safe and nutritious to sustain the day-to-day lives of millions of people and hence needs eternal vigilance to avoid related health hazards.

The students are expected that to read the laboratory procedures before class so that he or she may perform experiments more efficient and understand the reason for the results obtained. A unique aspect of this manual is inclusion of considerable background information on the relevant of Post Harvest Processing Technology with lab exercise to demonstrate these principles. Thus, students have all the relevant background information at hand to understand a lab exercise. In last we sincerely invite the readers to feel free and write to us expressing their appropriate opinions, critical comments and constructive suggestions.

This manual is unique in several aspects: it is an updated and comprehensive reference source, it contains topics not covered in similar other manual.

The author thanks concerned faculties, Research scholars, students and family members for supporting me to prepare this manual. I hope you enjoy and benefit from the fruits of their labor.

I know how hard it is to develop the content of a manual. However, I believe that the production of a professional manuscript of this nature is even more difficult. I thank the production team of New India Publication Agency, New Delhi to coordinate the entire project.

You are the best judge of the quality of this Manual.

Suresh Chandra

Contents

Concept of Skill Enhancement Course

Concept

Education is at the heart of both personal and community development; its mission is to enable each of us, without exception, to develop all our talents to the full and to realize our creative potential, including responsibility for our own lives and achievement of our personal aim. There is, therefore, every reason to place renewed emphasis on the moral and cultural dimensions, apart from what is taught presently, enabling each person to grasp the individuality of other people and to understand the world's progression towards an uncertain future. The 21st century is showing all signs of providing unprecedented means for communication and for the circulation and storage of information. Therefore, it will impose on education two demands which at first sight may appear contradictory. Education must transmit, efficiently and on a massive scale, an increasing amount of constantly evolving knowledge and know how adapted to a knowledge driven civilization, because this forms the basis of the skills of the future. At the same time, it must find and mark the reference points that will make it possible, on the one hand, for people not to be overwhelmed by the flow of information, invading the public and private domains and, on the other hand, to keep the development of individuals and communities as its end in view. Education must, as it were, simultaneously provide maps of a complex world in constant turmoil and the compass that will enable people to find their way in it. In this view of the future, traditional responses to the demand for education that are essentially quantitatively, and knowledge based are no longer appropriate. It is not enough to supply each student with a store of knowledge to be drawn on from then on. Instead, everyone must be equipped to seize learning opportunities throughout life, both to broaden her or his knowledge, skills, and attitudes, and to adapt to a changing, complex and interdependent world.

The restructured undergraduate curricula are designed to enable the students to take up entrepreneurship as a career path. As per NEP- 2020, the curricula in all the disciplines of Agricultural education have been refined and fine-tuned with intensive focus on choice-based skill enhancement programs. Skill enhancement courses are included in different modes as follows:- Skill enhancement courses in the 1st year and 2nd year as part of the course programs;. Internship for exit programs after 1st year/ 2nd year; and · Advanced skill enhancement through Student READY programs such as in-plant training/ internship/ projects in 4th year. The skills acquired must make the students proactive, pioneering, prospect oriented during their internship or industrial attachment to serve as apprentices in the relevant field. It will empower them to grasp viable avenues of self-employment and entrepreneurship along with diversified career options in different facets of related domains.

Skill enhancement courses are essential requirements for any programme (UG-Certificate/ UG Diploma or Degree). These will be choice based; student can choose the areas of skill enhancement from a bouquet of skill enhancement modules offered by the parent institution. The institutions will develop capabilities for offering such courses. An institution is at liberty to (and in fact, it should) work in partnership with capable organizations/ companies/ NGOs/ progressive entrepreneurs/ farmers for running various skill enhancement programs. The University/ HAEIs may also formulate and offer courses in any other areas as identified by it, based on institutional expertise/ capabilities/ resources. In addition, the skill enhancement courses suggested by the UGC, may also be offered.

An Overview
Indian Food Processing Industries

India has the second largest producer of food with arable land of 161 million hectares and has the highest acreage under irrigation. Next to China, India ranks second largest food producer in the world and has the potential to immerge the biggest with its food and agricultural sector. India accounts for less than 1.5% of international food trade despite being one of the world's major food producers, which indicates huge potential for both investors and exporters. Food and food products are the biggest consumption category in India, with spending on food accounting for nearly 21% of India's GDP and with a market size of $181 billion. The Indian domestic food market is expected to grow by nearly 40% of the current market size to $258 billion by 2015 and $344 billion by 2025 (World of Food India, 2011; Merchant, 2008). India's agricultural base is quite strong but wastage is very high and processing of food products is very low. While processing of food to consumable standards are at levels of up to 80% in some developed countries, the overall processing level in India has recently reached 10%. Therefore, India's food processing sector comparatively is small and its share in exports of processed food in world trade has remained at about 1.5 percent or $3.2 billion (Bhuyan, 2010).

The agro food processing industry is one of the largest in India, employs around 18% of the country's industrial work force and is ranked fifth in terms of production, consumption, export and expected growth (Merchant, 2008). India also produces a variety of temperate to tropical fruits, vegetables and other food products. Processing of food products plays an important role in the conservation and effective utilization of fruits and vegetables. India's strong agricultural base, variety of climatic zones and accelerating economic growth holds significant potential for food processing industry that provides a strong link between agriculture and consumers. The purpose of this paper is to examine trends and status of food processing industry in India. The paper also identifies the constraints/problems encountered and discusses challenges slowing down the growth of this sector. At the end, the paper examines opportunities and offers some feasible suggestions for continuous growth of the industry. Strength, Weakness, Opportunities and Threats analysis is used to highlight opportunities and threats facing the food processing industry and consider strategies to develop markets worldwide for processed food products.

Indian agri-business: Key facts

- Varied agro climatic zones
- 2nd largest arable land (161 million hectares) in the World
- Largest irrigated land (55 million hectares) in the World
- Largest producer of wheat (105 million tones), accounting for nearly 15% of global wheat production
- Largest producer of pulses (24 million tones), accounting for nearly 21% of global pulse production
- Largest producer of milk (230.58 million tons), accounting for nearly 17% of global milk production.
- Largest producer and exporter of spices
- 2nd largest producer of tea, accounting for nearly 28% of the global tea production
- 2nd largest producer of rice (137.8 million tons), accounting for nearly 26% of global rice production
- Largest exporter of the world's best basmati rice
- 2nd largest producer of fruits (112.73 million tons) and vegetables (204.96 million tons)
- 2nd largest producer of sugarcane (500 million tons), accounting for nearly 18.7% of the global sugarcane production
- 3rd largest producer of coarse grains (50 million tons), including maize, accounting for nearly 4% of the global coarse grain production
- 3rd largest producer of edible oilseeds (36.5 million tons), accounting for nearly 7% of the global edible oilseed production.

The Food Processing Sector

Food processing involves any type of value addition to the agricultural produce starting at the post harvest level. It includes even primary processing like grading, sorting, cutting, seeding, shelling packaging etc. The food processing sector comprises six major segments:

Grain Processing: The country's current food grain production (including rice, jowar, bajra, maize, ragi, wheat, barley, gram and pulses) has been put at 225 million tones a year. Food processing industries play a crucial role in reducing post-harvest losses. Since most operations of this industry are rural based, it has the potential to generate high employment at low investment. Promotion of food processing also helps in energy conservation by reducing energy wastages in home cooking. Grain processing, with a share of 40%, is the biggest component of the food sector. Its basic feature is pre-dominance of the primary processing sector, sharing 96% of the total value, with the secondary and tertiary sectors adding about 4%. This area needs to be viewed as a high growth potential area. Indian Basmati rice commands a premium in the international market.

Fruits and Vegetables: Horticultural crops in India are currently grown on 12 million hectares representing 7% of the country's total cropped area. Annual horticultural production is estimated at 100 million metric tones, which is over 18% of India's gross agricultural output. India is the world's second largest producer of fruits and vegetables.

Milk and Milk Products: India has the largest livestock population with milk cows and buffaloes being its main constituent. India is the world's largest milk producer (230.8 million tones annually) and the dairy industry has been recognized the world over as its most successful development programme going by FAO estimates, while world milk production fell by 2% in the last three years, the Indian production galloped by 4%. While consumption of liquid milk accounts for 46% of the total production, the rest is converted into milk products.

Meat and Poultry: India has the world's largest livestock population 535.78 million, accounting for 50% of buffaloes and 1/6th of the goat population. Such a large population represents a challenge to retain existing productivity traits by application of modern science and technology. Rigorous efforts are being made to improve the condition of livestock by providing basic infrastructure and latest technology. FAO has estimated the existing production of meat and poultry products at 4.42 million tonnes. Only 11% of the buffalo population, 6% of cattle, 33% of sheep and 38% of the goat population is culled for meat. At present, only a small percentage of the meat produced is converted into value added products and most meat is purchased by consumers in the fresh/ frozen form for conversion into products at home, restaurants, etc. Maximum conversion takes place in pork products. Some branded products like Venky's and Godrej's Real chicken are however becoming popular in the domestic market.

Fisheries and Marine Food: India boasts of the seventh largest marine landing base in the world with an extensive 7,500 km coastline and an Exclusive Economic Zone (EEZ) of 2 million sq km, largely untapped, and a 29,000 km stretch of rivers and canals, 145 million hectares of reservoirs and 0.75 million hectares of tanks and ponds. India is largest fish production country contributing 8% to the global fish production and ranks second in aquaculture production. The fish production in 2021-22 is 16.24 million tonnes comprising of marine fish production of 4.12 million tones and 12.12 million tonnes from aquaculture. There is also a good scope to improve fish harvest from inland waters which, at present is 13.113 million tonnes. Besides, the fish potential in aquaculture and shrimp farming has also largely remained untapped.

Alcoholic Beverages: Liquor made in India is categorised as beer, country liquor and Indian Made Foreign Liquor (IMFL). Country liquor is made from a variety of raw materials and has different names in different parts of the country. IMFL production comprises wine, vodka, whisky, gin, rum, brandy, etc. Pre-mixed drinks like gin and lime, rum and cola are being introduced in India now. Draught beer is another recent introduction and has done well where introduced. Canned beer is also a recent introduction. There are 512 breweries with a lincensed capacity of 4200 million litres alcohol per annum. The brewery and distillery industry industry is estimated at a market size of over USD 52 billion and expected to reach USD 70 billion by 2026. The Indian beer market is dominated by strong beers (>5% alcohol by volume), which account for 70% of the total beer industry. The premium beer market is a mere 5% of the total but this segment is rapidly expanding touching a growth rate of between 35-40%.

Major Challenges for the Indian Food Industry

Food-processing industry is facing constraints like non-availability of adequate infrastructural facilities, lack of adequate quality control & testing infrastructure, inefficient supply chain, and seasonality of raw material, high inventory carrying cost, high taxation, high packaging cost, affordability and cultural preference of fresh food. Unprocessed foods are prone to spoilage by biochemical processes, microbial attack and infestation. Good processing techniques, packaging, transportation and storage can play an important role in reducing spoilage and extending shelf life. The challenge is to retain the nutritional value, aroma, flavor and texture of foods, and presenting them in near natural form with added conveniences. Processed foods need to be offered to the consumer in hygienic and attractive packaging, and at low incremental costs. Major Challenges for the Indian Food Processing Industry are:

- **Inadequate Infrastructure Facilities:** The inadequate support infrastructure which is the biggest bottleneck in expanding the food processing sector, in terms of both investment and exports includes: long and fragmented supply chain, inadequate cold storage and warehousing facilities, road, rail and port infrastructure.
- **Absence of Comprehensive National Level Policy on Food Processing Sector:** The food processing sector is governed by statues rather than a single comprehensive policy on food processing. India urgently needs a national food processing policy which incorporates tax breaks

for the sector. The policy to be effective will have to be comprehensive & adopt a number of legislative, administrative & promotional measures.

- **Food Safety Laws & Inconsistency in State and Central Policies:** The Indian food regulations comprise various food policies that have been enacted at different points of time, and are under the ambit of various ministries of Government of India (GOI). Historically they were introduced to complement and supplement each other in achieving total food sufficiency, safety and quality. The result is that the food sector in India is governed by a number of different statutes rather than a single comprehensive enactment. This incremental approach has lead to incoherence and inconsistency in the food sector regulatory scenario.
- **Lack of Adequate Trained Manpower:** Many positive developments in the food processing sector have also resulted in the apprehension about the emerging skill shortages due to mismatch between the demand for specific skills and available supply. In fact, of late, shortage of skilled, semi-skilled and unskilled workers has emerged as a critical factor impacting the competitiveness of Indian food industry.

Apart from the above major challenges hampering the growth of sector constraints in raw material production, Consumer education on nutritional facts of processed foods, Low price-elasticity for processed food products Need for distribution network and cold chain, Backward-forward integration from farm to consumers, Development of marketing channels, Development of linkages between industry, government and institutions, lack of applied research etc as other major challenges for the growth of food processing sector. Indian food industry is gradually making an important mark in the global food arena as a large producer and exporter of agro food products. At present small players dominate the Indian food processing industry.

Constraints in Indian Food Processing Activities

Major constraints for the growth of the Indian food processing industry include the absence of adequate infrastructure, particularly rural road connectivity, inadequacy of information and marketing linkages, lack of electricity supply and the absence of cold chain systems. The cold chain capacity caters to less than 10 percent of the produce and within that facilities are so rudimentary that over 80 percent are only capable of handling potatoes. Maintaining the standards of quality is another major constraint and there are two aspects to it. First, there is poor infrastructure for storing raw food materials. The two main types of storages – the warehouses and the cold storages, lag in storage standards. The pests infest the grains sometimes due to lack of monitoring proper use of pesticides and proper ventilation. Similarly, the power outages result in suboptimal function of the cold-storages and the quality of food material in the cold storages becomes questionable. The second important aspect is having poor quality standards and control methods for implementing the quality standards for processing and packaging the processed foods. For example, vegetables may not be washed properly and processed into either 'ready to eat food' or packaged as 'cut and ready to cook' vegetables. High costs and low availability of credit remain a problem because even within the priority sector, lending by banks for agriculture, food processing receives only 4.5 per cent of the ear marked credit. The regulatory framework preventing farmers from directly marketing their produce, except through designated agricultural markets adds to cost and impairs flexibility. Packaging is usually poor but its cost is high and become unbearable for small producers. Another important constraint is the legal framework currently; food laws span nine ministries, comprising 13 central orders alone. In addition, states have their own control orders. In Table 2, SWOT analysis of Indian food processing industry is presented.

Table 2: SWOT Analysis of Agro-Processing Industry Infrastructure in India

Strengths	• Round the year availability of raw materials. • Social acceptability of agro-processing as important area and support from the central government. • Vast network of manufacturing facilities all over the country. • Vast domestic market.
Weaknesses	• High requirement of working capital. • Low availability of new reliable and better accuracy instruments and equipments Inadequate automation w.r.t. information management. • Remuneration less attractive for talent in comparison to contemporary disciplines. • Inadequately developed linkages between R&D labs and industry.
Opportunities	• Large crop and material base in the country due to agro-ecological variability offers vast potential for agro processing activities. • Integration of developments in contemporary technologies such as electronics, material science, computer, biotechnology etc. offer vast scope for rapid improvement and progress. • Opening of global markets may lead to export of our developed technologies and facilitate generation of additional income and employment opportunities.
Threats	• Large crop and material base in the country due to agro-ecological variability offers vast potential for agro processing activities. • Integration of developments in contemporary technologies such as electronics, material science, computer, biotechnology etc. offer vast scope for rapid improvement and progress. • Opening of global markets may lead to export of our developed technologies and facilitate generation of additional income and employment opportunities.

Reformation of Food Industry

i) Food safety and Standard Act, 2006

Thirteen different laws were applicable on the food and food processing sector till a Group of Ministers (GoM) was set up to rationalize and simplify the laws governing the industry in 2005. Thus, the Food Safety and Standard Act (FSSA) was formulated in 2006. The salient features of the act are:

- Food Safety and Standard Authority (regulating body of FSSA) will be aided by scientific panels and central advisory committee to lay down standards for food safety. These standards will include specifications for ingredients, contaminants, pesticide residue, biological hazards and labels.
- The law will be enforced through State Commissioners of Food Safety and local level officers.
- Everyone in the food sector is required to get a license or registration which would be issued by local authorities.
- Every distributor is required to be able to identify any food article to its manufacturer, and every seller is required to identify any food article to its distributor. Anyone in the sector should be able to initiate recall procedure if he finds that the food sold has violated specified standards.

ii) National Mission on Food Processing

The national mission on food processing was approved by the Government of India with an outlay of ` 1600 crore during the period 2012-17. The share of the central and the state government was ` 1250 crore and ` 350 crore respectively. The schemes taken up under the mission included:

- Scheme for technology upgradation / establishment / modernization of food processing industries
- Scheme for cold chain, value addition and preservation infrastructure for non-horticulture products
- Setting up/ modernization/ expansion of abattoirs
- Scheme for human resource development (HRD)
- Scheme for promotional activities
- Creating primary processing centres / collection centres in rural areas
- Modernization of meat shops
- Reefer vehicles
- Old Food parks

iii) Infrastructural Development

- Establishment of cold chain and provision of low cost pre-cooling facilities for farmers, entrepreneurs, traders and consumers would be encouraged and they will be trained to bring about attitude changes. Efforts will be made to disseminate market intelligence to enable farmers to fetch higher value for their produce.
- Efforts will be made to motivate farmers/industries to use insulated/refrigerated vehicles for transporting raw materials from the place of production/harvest to the point of consumption, to avoid wastage and quality deterioration.
- Establishment of cold storages and cold chain facilities would be encouraged.
- The interactive mesh between technology, economy, environment and society will be promoted for quicker development of agro-processing industries and to build up a substantial base for production of value-added products for domestic and export market with special emphasis on food safety and quality, taking into account all aspects of Total Quality Management (TQM) and Hazard Analysis and Critical Control Points (HACCP) to achieve international standards. Sustained R&D activities will be encouraged through recognised institutions having expertise in respective fields.

iv) Backward Linkage-Raw Material Supply

- The concept of backward linkage between farmers and industry would be promoted to encourage and enable farmers to grow products of appropriate quality. This will help the poorest of the poor farmers as well as marginal and medium farmers fetch appropriate and remunerative return for their produce. The scheme for providing assistance under the 9th Plan is already in operation and will be strengthened further to cover majority of farmers/producers. It would be ideal to dovetail this scheme with similar schemes offered by local bodies like Panchayats to provide the required fillip to this significant programme in the best interests of farmers and processors.
- The existing institutions like local bodies, cooperatives and self-help groups, which have been in operation for over four decades in different contexts, would be utilised to strengthen the backward linkage. This way the skill and expertise acquired by these institutions would be constructively used, while this mechanism would help quickly create the bridge of trust between farmers and processors. This would ensure

smooth supply of raw material to the processors and help the farmers (poor, marginal and big) in getting remunerative prices for their products. Thus, a complete network of farmers and processors will be created cutting across their status.

v) Forward Linkage-Marketing

- There is an urgent need to develop forward linkages for fresh and processed food. Presently, there are a large number of intermediaries operating between the farmers/processors and the consumers, resulting in high cost to the latter and low return to the former. The efforts to cut intermediaries would be made in such a way that the special skill and expertise required to operate the intermediate links in the system like transportation and market distribution are not jeopardized.
- To achieve this, attempts will be made to provide appropriate tax incentives and holidays for setting up food processing industries, taking care of expenses on market promotion and ancillary activities.

vi) Intensive and Extensive Awareness

- Extensive training will be provided to farmers and cooperatives in postharvest management of agro-produce to encourage creating pre-processing facilities near farms. Facilities may include provision for washing, fumigation, packaging, etc. Efforts will be made to encourage the setting up of agro-processing facilities as close to the area of production as possible to avoid wastages in transporting raw materials to far away places and to ensure increased value addition, specially for horticultural produce.
- Efforts will be made to improve general awareness about the advantages of consuming processed foods to stimulate domestic demand. Unfounded apprehensions about consuming processed food will also be removed.

Future Priorities

- Modernization of milling and processing of food grains, oilseeds, horticultural crop, animal products and fish.
- Development of processes and pilot plants for diversified uses of commercial crops, test marketing, industrial liaison and extension.
- Genetic manipulation for long shelf life, disease and environmental stress resistance cultivars.
- Modeling of post production system for optimization and on-line controls

- Environmental friendly storage pest management
- Optimal harvesting time and techniques.
- Post harvest treatments to increase shelf life storability (thermal, ultra-violet, irradiation, controlled atmosphere, chemicals).
- Energy conservation and efficiency in refrigeration and air-conditioning system.
- Standardization of CA storage and MAP for tropical fruits, ornamentals, planting material, fresh pack and minimal process produce.
- Microprocessor controlled processing and storage
- CA storage during transportation
- Humid forced aeration and cooling
- Establishing impact and vibration norms to prevent bruising and mechanical damage during sorting and packaging.
- Quantitative non-destructive measuring of quality and maturity of produce
- Environmental friendly packaging of produce and processed products.
- Study of consumer and market quality preferences in domestic and export market.
- Technological and management innovation to improve return on investment of post harvest enterprises.
- Basic research on post harvest physiology, senescence, ripening, respiration, ethylene effect, chilling, fermentation, superficial browning.
- Energy audits of post production systems and energy conservation measures.
- Study and application of supercritical CO_2 extraction of high value oil, oleoresin and caffeine extraction.
- Processing and utilization of underutilized plants, crop residues, byproducts and waste into high value added products of food, feed, fuel, fiber, pesticides and pharmaceuticals value.

Conclusion

The food processing sector is critical to India's development, for it establishes a vital linkage and synergy between the two pillars of the economy-Industry and Agriculture. India is the world's second largest producer of food and holds the potential to acquire the numero uno status with sustained efforts. The enormous growth potential of this sector can be understood from the fact that food production in the country is expected to double in the next

10 years, while the consumption of value-added food products will also correspondingly grow. The growth of this industry will bring immense benefits to the economy, raising agricultural yields, enhancing productivity, creating employment and raising life-standards of a large number of people across the country, especially those in rural areas. Major challenges faced by the Indian food processing industry include: educating consumers that processed foods can be more nutritious; dealing with low price elasticity for processed food products; need for distribution network; development of marketing channels; streamlining of food laws; improving food quality standards and strengthening food testing network; strengthening institutional framework to develop manpower for improving R&D capabilities to address global challenges. These challenges must be addressed to achieve full potential of the Indian food processing industry. The complexity of multiple administering authorities for food processing enterprises is also required to be simplified by developing anintegrated and unified system.

Bibliography

Bhuyan, A (2010) "India's Food industry on the Path of High Growth" Indo-Asian News Service.

Merchant, A (2008) India- Food Processing Industry OSEC Business Network land. www.osec.ch

Negi, Saurav and Anand, Neeraj (2015). Issues and challenges in the supply chain of fruits & vegetables sector in India: A review. International Journal of Managing Value and Supply Chains, 6(2): 47-62.

Prabhavathi, Y., Krishna Kishore, N.T. and Ramesh Kumar, M. (2014). Problems and Changing Needs of Consumers in Fast Food Industry: The Indian Perspective. International Journal of Scientific and Research Publications, 4(2): 1-4.

Singh, S. P., Tegegne, F. and Ekenem, E. (2012). The Food Processing Industry in India: Challenges and Opportunities. Journal of Food Distribution Research, 43(1): 81-89.

Sridhara Murthy, T.M. and Yogesh, M.S. (2014). An overview of Food Processing Industry in India – Challenges and Opportunities. Online International Interdisciplinary Research Journal, IV(V): 187-193.

Suresh Chandra, Durvesh Kumari, Samsher and Vishal Kumar (2018). Crop Process Engineering. The Jain brothers, Karol Bagh, New Delhi.

Experiments

Experiment 1

To Visit the Food Processing Unit and Write About Machineries

Objective: *To visit the Food Processing Unit and write about Machineries.*

Description

Experiment 2

To Visit the Agro Processing Centre and Write About Machineries

Objective: *To visit the Agro Processing Centre and write about Machineries.*

Description:

Experiment 3

Sensory Evaluation of Any Food Products

Objective: *Sensory evaluation of any food products.*

Material required

Food products, testing panel and sensory table.

Principles

The quality of foods can be defined as the 'composite' of those 'characteristics' that differentiate individual unit of a product and have significance in determining the degree of acceptability of that unit by the consumer. The overall quality refers to technological, physiological, chemical, microbiological, nutritional; and sensory parameters to achieve the wholesome food. Sensory quality attributes such as taste, colour, texture, flavour and overall quality of any food products were evaluated by 9- point Hedonic rating test is used for the purpose of evaluation. This test measures the consumer's acceptability.

Procedure

A trained panel consisting of 10 judges of different age groups having different eating habits is constituted to evaluate the quality through properly planned experiments. Different attributes viz. colour, texture, flavour and overall quality are rated on the basis of 9-point Hedonic Scale ranging is as Like extremely (9), Like very much (8), Like moderately (7), Like slightly (6), Neither like nor dislike (5), Dislike slightly (4), Dislike moderately (3), Dislike very much (2) and Dislike extremely (1).

Table Sensory

Samples	Taste	Colour	Flavour	Texture	Overall acceptability
1					
2					
3					
Total					
Average					

Result: Overall acceptability score =

Remark: Product is acceptable or unacceptable

Experiment 4

Development of Fruit Jam and Judging the TSS

Objective: *Development of fruit jam and judging the TSS.*

Material required

Fruit (apple), sugar, water, grinder, pan, heating system, preservative, glass jar and Refractometer.

Principle

Jam: Jam is a product made by boiling the fruit pulp with a sufficient quantity of sugar to reasonable consistency, firm enough to hold fruit tissues in position. Apple, pear, sapota, apricot, carrot etc. are used for preparation of fruit jam. Jam contains 0.5-0.6 % acid and invert sugar should not be more than 40 %.

Total soluble solids (TSS)

The TSS value is defined as the amount of sugar and soluble minerals present in fruits and vegetables. A refractometer is based on the principle of total refraction. This is done by placing a few drops of distilled water on the prism in the specimen chamber of the back the cover. The distilled water reading should be zero. If it is not so, it should be set to zero with the scale correction knob. The specimen chamber is now cleaned with muslin cloth or tissue paper. For determining the TSS, a drop of sample (jam, juice and syrup, etc.) is placed on the prism and the percentage of dry substance in it read directly on scale in refractometer.

Recipes

Fruits/Vegetables	**Ingredients for one kg pulp**		
	Sugar, kg	**Citric acid, g**	**Water, ml**
Apple	0.75	2.0	100
Carrot	0.75	2.5	200
Mango	0.75	1.5	50
Sapota	0.75	3.0	150

Process chart of Jam

Ripe firm fruits (Apple)

↓

Washing and grading

↓

Peeling

↓

Pulping → Seed

↓

Addition of sugar → Pulp ← Add water if needed

↓

Boiling with continuous stirring

↓

Addition of citric acid and KMS

↓

Judging of end point for further cooking up to 105^0C temperature or 68-70^0Brix TSS or by sheet test

↓

Filling hot in sterilized bottles

↓

Cooling

↓

Capping

↓

Waxing

↓

Storage at ambient conditions

Result

TSS = ……….^{0}Brix.

Taste = Like / Dislike

Experiment 5

Development of Fruit Jelly and Judging the TSS

Objective: *Development of fruit jelly and judging the TSS.*

Material required

Fruits, sugar, water, grinder, pan, heating system, preservative, glass jar and Refractometer.

Principle

Jelly: Jelly is a semi-solid product prepared by boiling a clear, strained solution of pectin containing fruit extract, free from pulp, after the addition of sugar and acid. A perfect jelly should be transparent, well set, but not too stiff, and should have the original flavour of the fruit. It should be of attractive colour and keep its shape when removed from the mould. Pectin (0.5-1.0%), acid (0.5-0.75%), sugar (65%), and water are the four essential ingredients. Pectin test and determination of end-point of jelly formation are very important for the quality of the jelly. Some recipes for jelly making are given below:

Fruits/Vegetables	Ingredients for one kg pulp	
	Sugar, kg	Citric acid, g
Guava	0.75	3.0
Sour apple	0.75-1.00	2.0
Gooseberry	0.80	-
Loquat	0.80	2.0
Papaya	0.75	3.0

Process chart of Jelly

Ripe firm fruits (Guava)

↓

Washing and grading

↓

Peeling

↓

Juice extraction → Seed

↓

Juice → Addition of sugar

↓

Boiling with continuous stirring

↓

Addition of citric acid, KMS and pectin

↓

Judging of end point for further cooking upto 105^{0}C temperature or 70-72^{0}Brix TSS or by sheet test

↓

Filling hot in sterilized bottles

↓

Cooling

↓

Capping

↓

Waxing

↓

Storage at ambient conditions

Resul:

TSS = ...^{0}Brix.

Taste = Like / Dislike

Experiment 6

Development of RTS Beverage From Seasonal Fruits

Objective: *Development of RTS beverage from seasonal fruits.*

Material required

Fruit, juicer-mixer-grinder, preservatives, sugar, refractometer, pan and knife etc.

Principle

Ready To Serve(RTS): This a type of fruit beverage which contain at least 10% fruit juice and 10% total soluble solid besides about 0.3% acid. This is not diluted before serving.

Before undertaking the preparation of beverage, it is necessary to know techniques of extraction of pulp/juice from various fruits used for RTS.

Recipe of RTS

Types of RTS	Juice/pulp (litre)	Sugar (Kg)	Water (litre)	Citric acid (g)
Bael	1.0	1.2	7.7	2.8
Lemon/Lime	1.0	1.3	8.2	-
Guava	1.0	1.25	7.7	2.8
Aonla blend (10 part aonla juice : 2 part lime juice : 1 part ginger juice)	1.3	1.6	10.0	2.2
Mango	1.0	1.25	7.7	2.8
Ginger	2.5	1.3	8.4	3.0

Process Chart of RTS Fruit Beverage

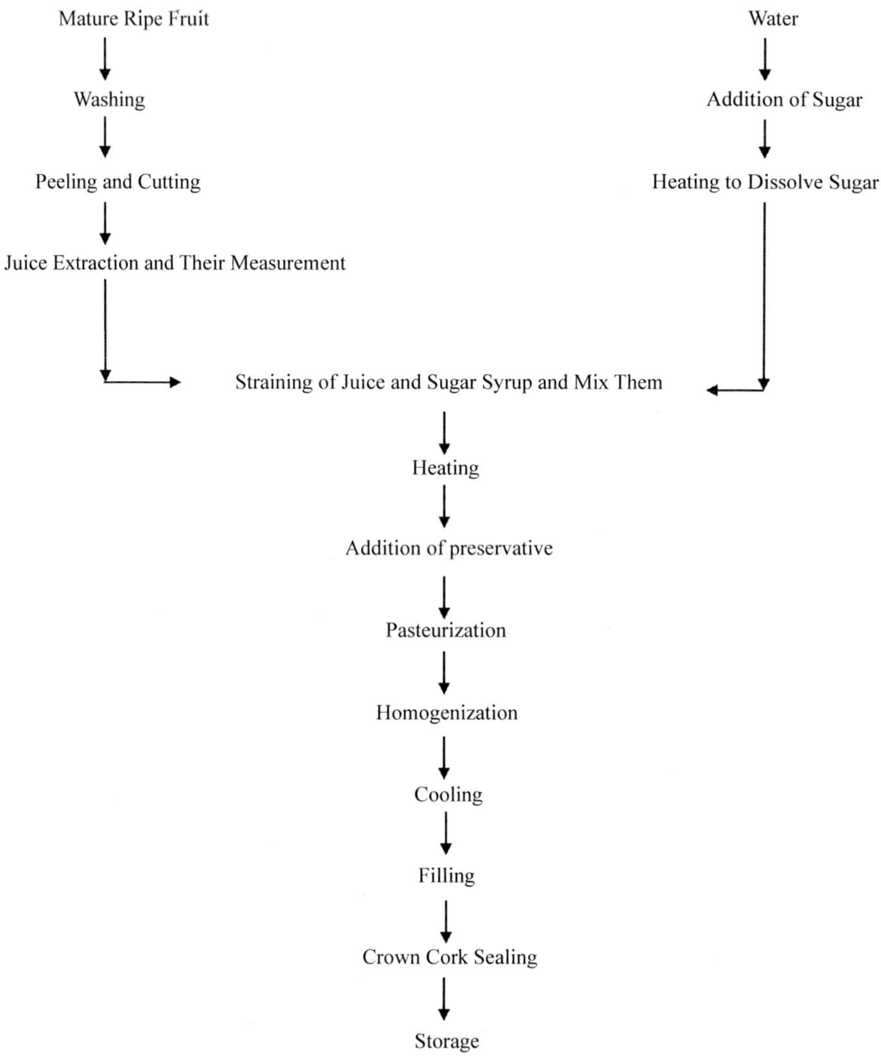

Result: TSS:………………… Taste: ………………

Experiment 7

Development of Fruit Squash and Determine the Taste and Final TSS

Objective: *Development of fruit Squash and determine the taste and final TSS.*

Material required

Fruit, juicer-mixer, preservatives, sugar, pan and knife etc.

Principle

Squash: A fruit juice squash essentially consists of juice containing moderate quantity of fruit pulp to which cane sugar is added for sweetening e.g. orange squash, plum squash. This is a type of fruit beverage containing at least 25 percent fruit juice/pulp and 40-50 percent total soluble solids, commercially. It also contains about 1.0 percent acid and 350 ppm SO_2 or 600 ppm sodium benzoate. It is diluted before serving.

Mango, orange and pineapple are used for making squash commercially. It can also be prepared from lemon, lime, guava, pear, apricot, muskmelon, papaya etc., using potassium metabisulphate (KMS) as preservative, or from jamun, passion fruit, peach, phalsa, mulberry, raspberry, straw berry, grapefruit etc., with sodium benzoate as preservative.

Types of squash	Juice/pulp (litre)	Sugar (kg)	Water (litre)	Preservatives (g)		
				Citric acid	KMS	Sodium benzoate
Apricot	1.0	1.8	1.2	2.5 g	1.5 g	-
Litchi	1.0	2.0	1.0	-	-	3.0 g
Lime/Lemon/Orange	1.0	1.8	1.2	-	3.0 g	-
Mango	1.0	1.5-2.0	1.0-1.2	2.5-3.0	2.0 g	-
Bael	1.0	1.80	1.0	2.5	2.5	-
Pineapple	1.0	1.75	1.0	2.0	1.9	-
Jamun	1.0	1.80	1.0	1.5	-	3.0
Guava	1.0	1.80	1.0	2.0	2.0	-

Process Chart of Fruit Squash

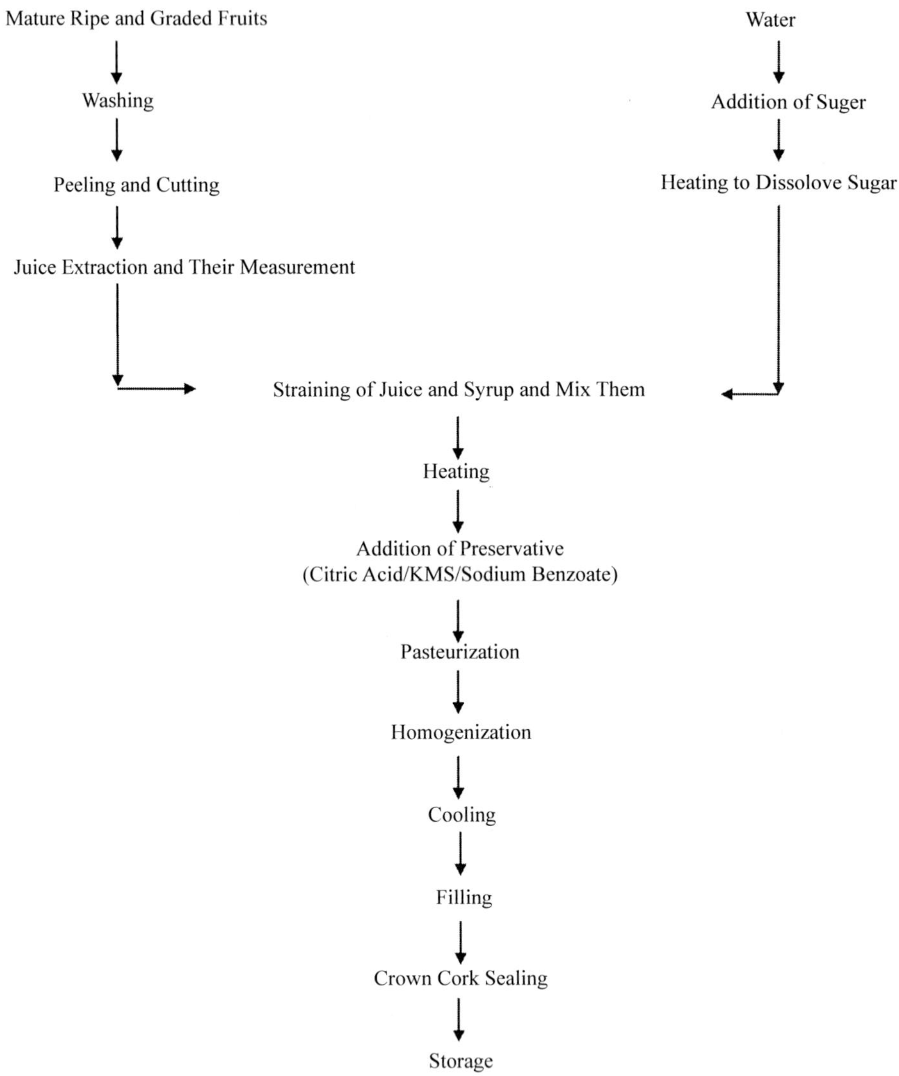

Result: TSS: .... Taste: ....

Experiment 8

Development of Fruit Nectar and Determine the TSS and Taste

Objective: *Development of fruit Nectar and determine the TSS and Taste.*

Material required

Fruit, juicer, preservatives, sugar, pan and knife etc.

Principle

Nectar: The term 'fruit nectar' is used to designate pulpy fruit juice blended with sugar syrup and citric acid to produce a ready-to-drink beverage. In this fruit beverage contains at least 20% juice/pulp and 15% total soluble solid and also 0.3% acid. It is not diluted before serving. Many fruit juices are either too acidic or strongly flavoured; these cannot be pleasant without dilution of blending or both. These strong, tart juices are delicious after dilution with thin syrup or blend juice e.g. apricot, guava and plum juice. Out of the tropical fruits, nectars made from banana, guava, mango and papaya are quite popular. The product is sealed in container or glass bottles and processed by heat to prevent spoilage. The fruit ingredients may be puree, pulp, juice or concentrated equivalent to 40 percent of the finished products. Apples, cherries and pineapples are used in combination with others fruits. The sweeteners are sugar, invert sugar syrup, dried corn syrup, glucose syrup and the optional ingredients are lemon juice, citric, malic and fumaric acids as acidifier and ascorbic acid as antioxidants (150 ppm) can be used.

Types of nectar	Juice/pulp (litre)	Sugar (kg)	Water (litre)	Citric acid (g)
Mango	1.0	0.60	3.3	1.3
Guava	1.0	0.60	3.3	1.3
Bael	1.0	0.65	3.3	1.5
Aonla blend (10 part aonla juice: 2 part lime juice: 1 part ginger juice)	2.3	1.35	7.8	0.4

Process Chart of Fruit Nectar

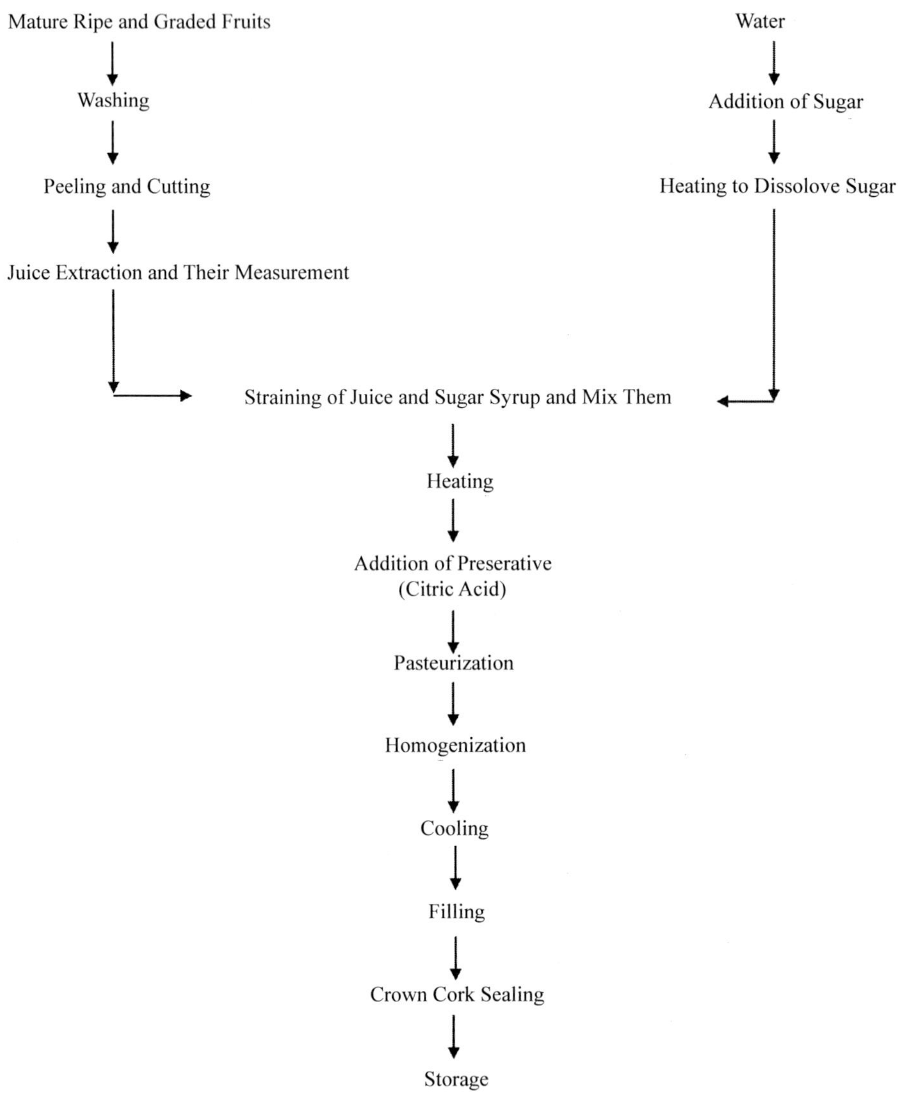

Result: TSS: .... Taste: ....

Experiment 9

Development of Tomato Sauce/Ketchup and Determine the TSS & Taste

Objective: *Development of tomato sauce/ketchup and determine the TSS & Taste.*

Material required

Tomato, spices, glass bottles, pan, preservatives and knife etc.

Principle

Chutney and sauces are important food products prepared both in homes as well as commercially in India. They improve the digestion and are good appetizers. Tomatoes are a good source of vitamin C. Fresh tomatoes are very refreshing and appetizing but can not be stored for a long period. Often they are sold at distress prices during the peak harvest season and nearly 25% of the produce is spoiled due to mishandling and improper processing facilities. Such losses can be reduced by converting fresh tomatoes in to delicious product like paste, puree, juice, ketchup/sauce and powder. Generally, the sugar content in ketchup/sauce varies from 10-26%. The salt content of the product should be 1.3 to 3.4%. Good quality vinegar is essential for the preparation of high quality sauce/ketch up. It should contain 5.0-5.5% acetic acid or glacial acetic acid (100% acetic acid) and should be added when the product has thickened sufficiently. Tomato sauce/ketchup generally contains 1.25-1.50% acetic acid. In order to increase the viscosity and prevent the separation of pulp from clear juice, pectin can be added to extent of 0.1-0.2% pectin by weight of the finished product. Add 0.025% sodium benzoate as preservative to the product before bottling and then pasteurize.

Recipe:

Tomato pulp 1 kg, sugar 75 g, salt 10 g, onion (chopped) 50 g, ginger (chopped) 10 g, garlic (chopped) 5 g, red chilli powder 5 g, cinnamom, cardamom (large), aniseed, cumin, black pepper (powdered) 10 g each, clove (headless) 5 numbers, vinegar 25 ml or glacial acetic acid 5 ml and sodium benzoate 0.25 g per kg final product.

Process Chart of Tomato sauce/ketch up

↓

Fruits and/or vegetables

↓

Grading for ripeness and freshness

↓

Washing

↓

Cutting in pieces

↓

Crushing

↓

Pulp Extraction → Skin and seed

↓

Pulp cooking with spices ← Add of vinegar if need

↓

Bottling

↓

Sealing

↓

Thermal sterilization

↓

Cooling

↓

Labeling and coding

↓

Storage and marketing

Result: TSS: ………………… Taste: ………………

Experiment 10

Development of Mixed Pickle from Different Fruits and Vegetables

Objective: *Development of mixed pickle from different fruits and vegetables.*

Material required

Cauliflower, pan, knife spices, vinegar and oil etc.

Principle

Pickle is an edible product preserved and flavoured in a solution of common salt and vinegar. Spices and oil are also added. They add to palatability of a meal, aid in digestion and are good appetizers too. Pickles are prepared from fruit and vegetables like cauliflower, reddish, carrot, green chili, turnip, cabbage, onion, mango, jackfruit, aonla and lemon etc. Commercial preservation of many pickle relied upon conversion of fermentable carbohydrates by lactic acid bacteria to organic acids during bulk storage and/or the addition of sufficient amount of sugar, vinegar and other ingredients to the fully cured and packed products to preclude any microbial growth. Organic acids, oil, salt and spices all have antimicrobial properties at suitable concentration especially in combination which preserves the pickles. Pickles in brine such as or after fermentation however, need some amount of preservative or pasteurization to prevent the spoilage.

Recipe

• Cauliflower = 1 kg,	• Red chili, turmeric, black pepper, cinnamon, cardamom (large), cumin, aniseed (powdered) each = 15 g,
• Salt = 50 g,	
• Ginger = 25 g,	• Clove (headless) = 6 numbers,
• Onion = 50 g,	• Tamarind pulp = 50 g,
• Garlic = 10 g,	• Vinegar = 150 ml,
• Mustard (ground) = 50 g,	• Mustard oil = 400 ml

Process chart of mixed pickle

Cauliflower

↓

Washing

↓

Cutting into 2.0 - 2.5 cm pieces

↓

Keeping in sun for 2 hrs

↓

Frying all spices in a little edible oil

↓

Mixing pieces with fried spice

↓

Frying mixture for 5 minutes

↓

Cooling

↓

Making paste of vinegar and tamarind pulp

↓

Adding to cauliflower fried pieces

↓

Filling in glass jar

↓

Keeping in sun for a week

↓

Addition of oil after heating and cooling

↓

Storage

Result: Taste: ....

Experiment 11

Development of Biscuit or Cookies

Objective: *Development of Biscuit or Cookies.*

Materials required

Material : Wheat Flour (1000g), Shortening (350 g), Fine Sugar (400 g), Sodium bicarbonate (10 g), Dextrose solution (100 ml) i.e 25 g Dextrose hydrate in 500 ml water, Distilled water (50 ml), salt (8 g).

Equipments: Mixer, rolling pin, Tray, cookies cutter, baking oven etc.

Principle: Cookies are made from soft wheat are having high sugar and fat content. They contain relatively low moisture content and have fairy high shelf life. Biscuits are similar name used in Europe and UK. Cookies are chemically leavened. Sugar and shortening are creamed to form dough with flour and leavening agents. When cookies dough enters the oven, the fat starts melting and sugar is dissolved which allow the cookies to spread as a function of gravity. The leavening agents expand it in all direction and structure is stabilized with high temperature prevailing in the oven.

Procedure

1. Cream shortening, sugar, salt and soda at low speed for 3 min and scrap the sides.
2. Add water and dextrose solution and mix for 1 min at low speed, 1 min at medium speed. Add all the flour and mix for 2 min at low speed.
3. Place the dough on metal sheet and divided into 6 pieces. Take one and flatten slightly with hand using gauge stripes, roll to thickness using rolling pin.
4. Cut the dough with cookies cutter and discard the excess dough.
5. Cut similarly the other five cookies and bake at 180°C for 20-30 min.
6. Cool the cookies.

Outlook: The cookies should be uniform, small cracks over the top and the appearance should be attractive, shape should be symmetrical and baked properly from all sides. They should be burning spots and it should be crisp with tender bite.

Results: Observe the final product for colour, texture and taste.

Experiment 12

Development of Multigrain Cookies

Objective: *Development of Multigrain Cookies.*

Materials required

Material : Wheat Flour (550 g), Pearl millet Flour (50 g), Ragi Flour (50 g), Finger millet Flour (50 g), Barley Flour (50 g), Sorghum Flour (50 g), Foxtel Millet Flour (50 g), Maize Flour (50 g), Oat Flour (50 g), Besan (50 g) Shortening (350 g), Fine Sugar (400 g), Sodium bicarbonate (10 g), Distilled water (50 ml), salt (8 g).

Equipments: Mixer, rolling pin, Tray, cookies cutter, baking oven etc.

Principle: Cookies are made from soft wheat are having high sugar and fat content. They contain relatively low moisture content and have fairy high shelf life. Biscuits are similar name used in Europe and UK. Cookies are chemically leavened. Sugar and shortening are creamed to form dough with flour and leavening agents. When cookies dough enters the oven, the fat starts melting and sugar is dissolved which allow the cookies to spread as a function of gravity. The leavening agents expand it in all direction and structure is stabilized with high temperature prevailing in the oven.

Procedure

1. Cream shortening, sugar, salt and soda at low speed for 3 min and scrap the sides.
2. Add water and dextrose solution and mix for 1 min at low speed, 1 min at medium speed. Add all the flour and mix for 2 min at low speed.
3. Place the dough on metal sheet and divided into 6 pieces. Take one and flatten slightly with hand using gauge stripes, roll to thickness using rolling pin.
4. Cut the dough with cookies cutter and discard the excess dough.
5. Cut similarly the other five cookies and bake at 180°C for 20-30 min.
6. Cool the cookies.

Outlook: The cookies should be uniform, small cracks over the top and the appearance should be attractive, shape should be symmetrical and baked properly from all sides. They should be burning spots and it should be crisp with tender bite.

Results: Observe the final product for colour, texture and taste.

Experiment 13

Development of Rusk

Objective: *Development of Rusk.*

Materials required

Equipments: Mixer, rolling pin, Tray, cookies cutter, baking oven etc.

Raw Materials

I) Flour (80 g), Yeast (1.5-2.0g), Fine Sugar (1-2 g), Skim milk powder (1-2 g), water (56-58 g).

II) Flour (20 g), salt (1-1.5g), Fine Sugar (25-30g), Yeast (1-1.5g), Shortening (15-20 g), khaskhas/caraway seeds/Aniseeds (optional).

Principle: Rusk is a dried product having moisture content of 2-4 %. It is prepared by two step baking followed by drying.

Procedure

1. Mix the dough with ingredients I. Keep it for fermentation till breaks easily without stretching.
2. Put the sponge to mixer. Add salt, sugar, flour and optional ingredients and mix. In last, add shortening. The dough should be soft and very smooth.
3. Let the dough relax for about 30 min. cut pieces of dough weighing 100 g/450g.
4. Mould in along oblong shape and place in baking pans.
5. Proof the moulded dough to a desired height.
6. Bake at 210°C to a golden colour.
7. Cool the loaves overnight.
8. Cut the rusk of desired thickness. Arrange on baking sheets and toast at 190°C till these are crisp and golden brown in colour.
9. Cool and packaged.

Result/Evaluation: Observe the final product for colour, texture and taste.

Experiment 14

Development of Sugarcane Vinegar

Objective: *Development of Sugarcane Vinegar.*

Principle: Sugarcane is one of the main crops worldwide, and it has an important impact on environmental issues. Sugarcane is used in daily routine life in many ways like as vinegar, jiggery, juices etc. In India sugarcane crop is best cultivated in the west U.P. Sugarcane original vinegar drink with high nutritional quality was produced from fresh sugarcane juice using the yeast culture and acetic acid bacteria by fermentation techniques such as submerged alcoholic fermentation followed by acetic fermentation at room temperature. Refined sugarcane is the primary product of sugarcane juice, during its processing; various other valuable products are also obtained in an unrefined form such as brown sugar, molasses, jaggery and vinegar. Sugarcane juice is widely used in the treatment of jaundice, hemorrhage, dysuria, and other urinary disease. Nowadays sugarcane vinegar is also used in Indian kitchens commonly in pickles, salads, etc. Vinegar is extremely useful for human health including antimicrobial activity, blood pressure reduction, antioxidant activity, reduction in the effects of diabetes, and prevention of cardiovascular disease. Vinegar can be made from any carbohydrate source, amylaceous, or sugary substrate through two successive fermentations: alcoholic fermentation, which is carried out by means of yeasts, and acetic fermentation, with acetic bacteria as the protagonist.

Ingredients for Vinegar

- Sugarcane Juice - 1 litre
- Water - 500 ml
- Nails - 2
- Acidic Acid - 2 tbsp

Natural Vinegar (Sugarcane Vinegar)

- Take 1 litre of Sugarcane Juice in a Glass Jar,
- Add 2 non-rusted Nails in it.
- Place it in a warm place or cupboard for the 2 months.

- After 2 months, filter it using a cotton cloth and after Filtering it,
- Store in a Glass, Porcelain or food grade plastic bottle.
- Sugarcane Vinegar is ready to be used.

Take Fresh Raw Material of Sugarcane

↓

Diluted with Distilled Water

↓

Incubate at Room Temperature for 3-4 Week

↓

Filtered Through Four Layered Muslin Cloth

↓

Repeat this Step 2-4 Times Until the Supernatant was Obtained

↓

After 3-4 Months Good Smell of Vinegar was Produced

Suggestions

- Do not use the metal strainer for the filtration process.
- If you have a strainer with plastic wires then, you can use it.
- The bottle which we will use for it should be of Glass, Porcelain or food grade plastic bottle.
- If you don't have RO water then, you can use water after boiling it.

Experiment 15

Preparation of Turmeric Powder

Objective: *Preparation of Turmeric powder*

Turmeric (Curcuma longa L), the ancient and sacred spice of India known as 'Indian saffron' is an important commercial spice crop grown in India. It is used in diversified forms as a condiment, flavouring and colouring agent and as a principal ingredient in Indian culinary as curry powder. It has anti-cancer and anti-viral activities and hence finds use in the drug industry and cosmetic industry. 'Kum-kum', popular with every house wife, is also a by-product of turmeric. It finds a place in offerings on religious and ceremonial occasions. A type of starch is also being extractcd from a particular type of turmeric. The increasing demand for natural products as food additives makes turmeric as ideal produce as a food colourant. India is the largest producer, consumer and exporter of turmeric in the world.

Harvesting

Turmeric is harvested when the plants are between 7 and 10 months of age, when the stems and leaves start to dry out and die back. The whole plant is removed from the ground, taking care not to cut or bruise the rhizomes.

Sweating

The leaves are removed from the plant and the roots carefully washed to remove soil. Any leaf scales and long roots are trimmed off. The side (lateral) branches (which are known as the fingers) of the rhizomes are removed from the main central bulb (known as the mother). The mothers and fingers are heaped separately, covered in leaves and left to sweat for one day. The 'mothers' are the preferred material for planting the following year.

Curing

Before drying, the turmeric rhizomes have to be cured. This involves boiling the roots to soften them and remove the raw odour. After curing, the starch is gelatinised, which reduces the drying time required, and the colour is uniformly distributed throughout the rhizome.

The specifications for curing turmeric vary from different places. The Indian Institute of Spice Research and the Agricultural Research Centre recommend boiling in plain water for 45 minutes until froth appears at the surface and the typical turmeric aroma is released. Using this method, the colour will deteriorate if the rhizomes are boiled for too long. However, if not boiled for long enough, the rhizome will be brittle. The optimum stage is reached when the rhizomes are soft to touch or can be pierced by a blunt piece of wood.

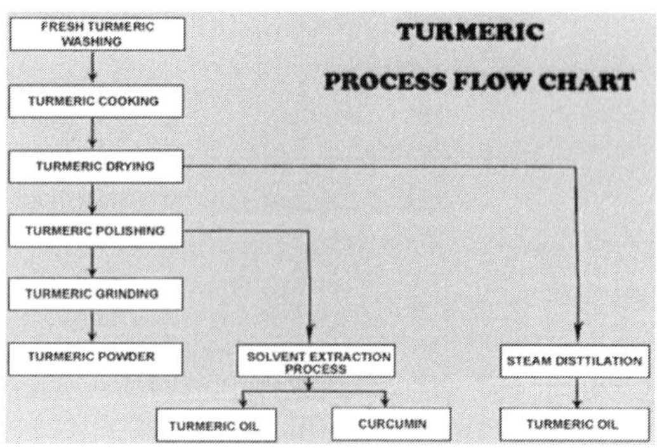

Other researchers recommend boiling the rhizomes in an alkaline solution, made from 0.05% to 1.0% sodium carbonate or lime (calcium carbonate). The alkaline water is thought to improve the final colour of the dried turmeric. Because the fingers and bulbs are of different size and thickness to each other, it is important that they are cured in different batches. The curing time is dependent on the age, variety and size and thickness of the rhizomes and varies from one to four or sometimes six hours. It is also affected by the size of the batch – smaller batches generally produce higher quality dried turmeric as the curing process is easier to control and monitor. It is recommended to use perforated containers with a capacity of 50-75kg. The perforated containers are immersed in the boiling water for the required cooking time. This method allows the boiling water to be used for more than one batch of turmeric.

The best time for curing is two to three days after harvest. If it is delayed after this then the rhizomes may start to deteriorate. The benefits of curing turmeric include the following:

- Reduction of drying time
- More even colour distribution throughout the rhizome
- A more attractive (not wrinkled) product that is easier to polish
- Sterilization of the rhizomes before drying.

Drying

The rhizomes are sliced before drying to reduce the drying time and improve the quality of the final product (it is easier to achieve a lower final moisture content in small pieces of rhizome without spoiling the appearance of the product). The rhizomes are traditionally sliced by hand, but there are small machines available to carry out this process. It is a simple structure that contains a transmission system and two stainless steel circular blades. The machine is easy to build and maintain and can cut up to 120 kg turmeric per hour.

The cooked fingers or bulbs are dried until they have a final moisture content of 5-10%. An experienced turmeric processor will know when the rhizome is dry enough as the fingers will snap cleanly with a metallic sound. Traditionally the rhizome pieces are laid on clean concrete floors and dried in the sun. This method can take anything from 10 to 15 days, depending on the climate and the size of the rhizome pieces. It is important that the rhizome pieces are not placed in direct sunlight as this will cause the colour to fade. Using a mechanical drier will result in a better colour and a higher quality product.

There are several different types of mechanical drier that are suitable for drying turmeric. These include the tray drier, cross flow air tunnels, solar driers and cabinet driers. The optimum drying temperature is 60°C temperatures higher than this result in a darker coloured product.

Polishing and colouring

Dried turmeric has poor appearance and rough dull outer surface with scales and root bits. The appearance is improved by smoothening and polishing the outer surface by manual or mechanical power operated drums. Polishing is done till the recommended polish of 7-8% is achieved. Usually 5 to 8% of the weight of turmeric is the polishing wastage during full polishing and 2 to 3% during half polishing. Polishing of dried turmeric also helps in removing the wrinkles.

Manual polishing consists of rubbing the dried turmeric fingers on a hard surface. Manual polishing gives rough appearance and dull colour to the dried rhizome. Sometimes, undesirable colouring materials are added during polishing to improve the colour and appearance. But this is not recommended. In an improved method, polishing is done by using hand operated barrel or drum mounted on a central axis, the sides of which are made of expanded metal screen. When the drum filled with turmeric is rotated, polishing is effected by abrasion of the surface against each other as they roll inside the drum. The turmeric is also polished in power operated drums.

Large scale polishing units with capacity to polish 500 to 1000 kg per batch is used for polishing turmeric rhizomes at commercial units. It takes about 45-60 minutes per batch and about 4% is wasted as dust.

The colour of the processed turmeric influences the price of the produce. Hence, to obtain attractive product, turmeric powder is sprinkled during the last phase of polishing.

Grading

Quality specifications for turmeric are imposed by the importing country. They refer to the cleanliness of the product rather than on the eating quality. Bulk rhizomes are graded into fingers, bulbs and splits. Although Indian turmeric is considered to be the best in the world, about 90% of the total produce is consumed internally and only a small portion of the production is exported. Turmeric of commerce is described in three ways:

Fingers: These are the lateral branches or secondary 'daughter' rhizomes which are detached from the central rhizome before curing. Fingers usually range in size from 2.5 to 7.5 cm in length and may be over 1 cm in diameter.

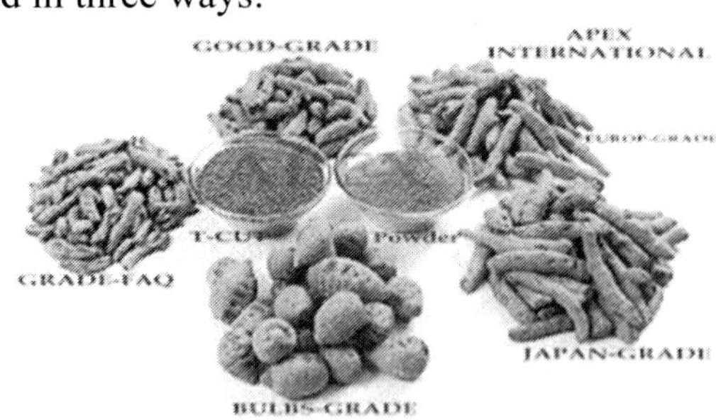

Bulbs: These are central 'mother' rhizomes, which are ovate in shape and are of shorter length and having larger diameter than the fingers.

Splits: Splits are the bulbs that have been split into halves or quarters to facilitate curing and subsequent drying.

Turmeric being a natural produce, is bound to gather contaminants during various stages of processing. The spice is also cleaned to remove such foreign materials. A sifter, destoner, and an air screen separator will help remove materials such as stones, dead insects, excreta, and other extraneous matter.

Cleaned and graded material is packed generally in new double burlap gunny bags and stored over wooden pallets in a cool, dry place protected from light. The stores should be clean and free from infestation of pests and harborage of rodents. It is not recommended to apply pesticides on the dried/polished turmeric to prevent storage pests.

Grinding

Grinding can be a method of adding value to a product. However, in general it is not advisable to grind spices as they become more vulnerable to spoilage. The flavour and aroma compounds are not stable and will quickly disappear from ground products. The storage life of ground spices is much less than for the whole spices. It is very difficult for the consumer to judge the quality of a ground spice. It is also very easy for unscrupulous processors to contaminate the ground spice by adding other material. Therefore, most consumers, from wholesalers to individual customers, prefer to buy whole spices.

Turmeric is one of the few spices that are usually purchased in a ground form. The whole rhizome pieces may be exported and then ground in the country of destination. Alternatively, the dried rhizomes may be ground at the place of origin. Grinding is a very simple process that involves cutting and crushing the rhizomes into small particles, then sifting it through a series of screens of different mesh size, to get a fine powder. There are a range of grinding mills available, both manual and powered, of different capacities and which work in different ways. The traditional way to grind would be between two stones.

The advantage of this method is that the turmeric does not get too hot during the grinding process. With some of the mechanical mills, such as a hammer mill, heat is generated during the grinding process, which can cause some of the volatile taste and aroma compounds to be lost. For higher quality ground turmeric, the grinding temperature should be kept as low as possible. After grinding the powder is sieved through different mesh screens until a uniform, fine powder is obtained.

Storage

The bulk rhizomes are stored in a cool and dry environment, away from direct sunlight. The bright colour of ground turmeric will fade when it is left in the light for a long period of time. Therefore, the packets should be stored in a cardboard box, away from the sunlight. The storage room should be clean, dry, cool and free from pests. Mosquito netting should be fitted on the windows to prevent pests and insects from entering the room. Strong smelling foods, detergents and paints should not be stored in the same room.

Preservation of seed rhizomes

Rhizomes for seed purpose are generally stored by heaping in well ventilated rooms and covered with turmeric leaves. The seed rhizomes can also be stored in pits with saw dust, sand along with leaves of Stychnos nux-vomica (Kanjiram). The pits are to be covered with wooden planks with one or two openings for aeration. The rhizomes are to be dipped in quinalphos (0.075%) solution for 20-30 minutes if scale infestations are observed and in mancozeb (0.3%) to avoid storage losses due to fungi.

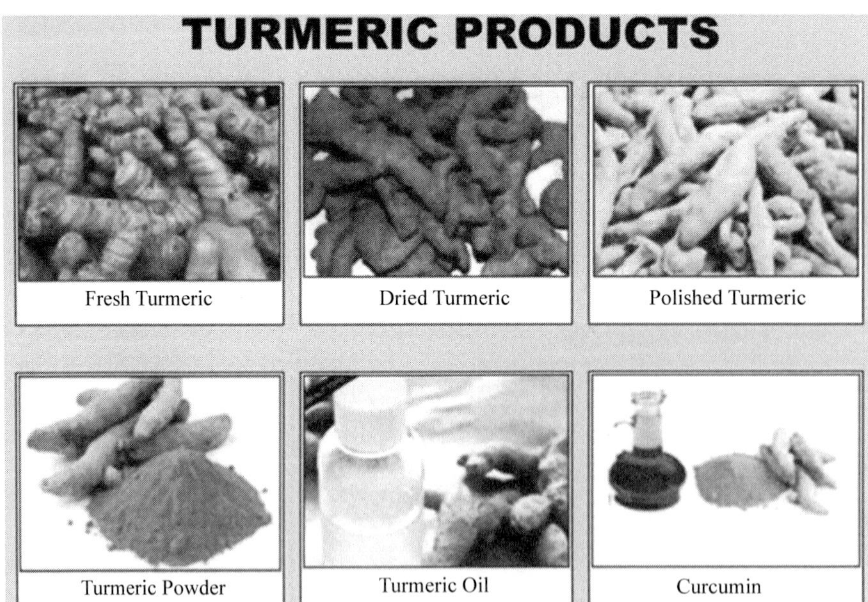

Source: KAU Agri-Infotech Portal

Experiment 16

Development of Potato Chips

Objective: *Development of Potato chips*

Potato is a perishable commodity and cannot be preserved longer in fresh form. Although for some time potatoes can be preserved by low temperature storage but this low temperature storage results in many undesirable biochemical changes in the quality of the potatoes. Another alternative for the preservation of potatoes is to conversion in the shelf stable value-added processed products. Further, value-addition and processing of potato will also ensure the availability of wholesome, safe, nutritious, and acceptable food to consumers throughout the year along with simultaneous reduction in post-harvest losses and profit to the farmers for their produce. Potatoes are processed into a great variety of products, including cooked potatoes, par-fried potato strips, French fries, potato chips, potato starch, potato granules, potato flakes and dehydrated diced potatoes and others. Globally, only two potato-based snack foods i.e., chips/wafers and French-fried potato are the main value-added products of potato, accounting for more than 42% of the potato based processed products. Such type of fried products is loved for convenience and their premium taste, flavour, and texture. With the technological advancements, now potato processing is highly industrialized and market driven. Like other processed products, the finished product quality of potato based value-added products severely depends upon the quality of raw potatoes used. Therefore, the major challenge in potato processing is to fulfil the tuber quality requirements specifically, tuber morphology, reducing sugar ($\leq$0.1%) and dry matter content ($\geq$20%). Alternatively, we can infer that all type of potato is not suitable for processing.

The crispy texture of potato chips is one of the most important quality indicators of the finished product, apart from colour, odour and flavour. Potato chips texture is often described in terms of crispness, hardness and crunchiness. This crispy/crunchy character is an important sensory characteristic on which consumers base their appreciation.

Material needed: Potato, chips cutter, edible oil, Electric fryer, spices, utensils etc.

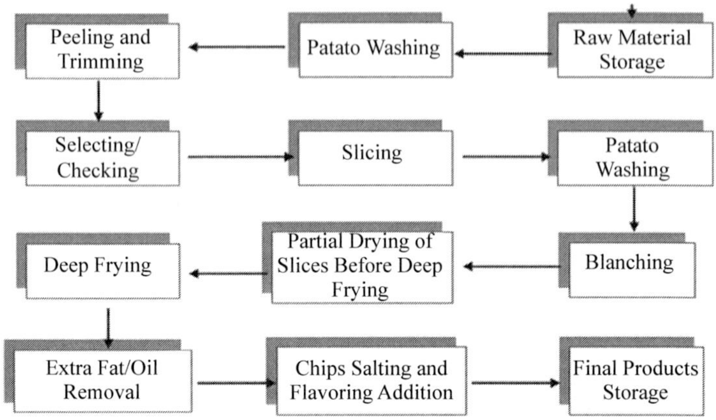

Description

1. Material Selection: Select fresh potatoes of high dry matter content, pure varieties, light eyes, regular shape, free from mildew rot, sprouting and insect infestations and other phenomena, 50~70mm in diameter.
2. Cleaning: Use roll cage washing machine to remove the dirt, sand and other debris.
3. Peeling: Use the mechanical friction peel mode, feed 30~40kg each time, the peeling time is determined by the degree of freshness of material, usually 3~8min. The peeled potatoes need to have net skin and smooth surface. Peeling time should not be too long or the potatoes will be excessive peeled. It will increase the loss rate of material.
4. Trimming: Remove the parts that unfit for human consumption which have buds, mildew and so on, and shape the irregular ones.
5. Potato Slicing: Feed the material into the potato slicing machine, feeding at a constant speed, the thickness should be 1~2mm.
6. Blanching: The potato slices should be immersed into water to blanch, in order to prevent browning in the air, and wash away the free starch of the surface to avoid the starch dissolving into the oil and shorten the life of frying oil.
7. Color Protecting: The temperature of the color protection solution should be 80~100°c and should destroy enzyme and improve the organizational structure. Usually the blanching time should be 1~2min. Besides that, it needs to add little additives.
8. Potato Chips Dehydrating: Spread the material stacked together on the subfeeder, and remove surface water, to avoid increasing the frying time and increase the oil content of the final product.

9. Potato Chips Frying: The potato slices should be put into the oil at a constant speed. The oil temperature should be at 180~200□. The chips should pass the frying pan in 2min. Palm oil is usually used in this process. It has good stability, long service life, good shortening properties, especially suitable for frying oil

The frying process is the most important in producing potato chips. It plays a decisive role in color, oil content, and water content. While, the frying temperature and time have a direct relationship with the dry matter content, the thickness of chips and the degree of blanching. Through lots of experiments, we find that frying at a low temperature makes the frying time longer, water loss slow and increase the oil content. It is better to fry at high temperature and short time, but the temperature should be less than 200°C. Because if the frying temperature is too high, the fat will decompose and produce fatty acids. Fatty acids dissolve metal ions which are a catalyst for decomposition of oil and cause oil rancid and shorten the period of use of oil. High temperature causes potato chips excessive sugar caramelizing reaction which affects seriously the color and luster of the product. So we make the frying temperature at 180~190°c and the frying time in 2min and use stainless steel fryer. To prolong the life of frying oil can add antioxidants BHA and BHT and the total amount of antioxidants shall not exceed 0.12/kg fat. The oil content is related to many factors of the product. The higher the dry matter content of the product, the lower the oil content. Because the higher the dry matter content of the raw material, thc more dense the organization, the lower the water content, and the shorter the drying time. Second the oil content is inversely proportional to the slice thickness. If the slice is too thin, it is easy to fry excessively and the oil content increase. Generally, the thickness is no less than 1mm.

10. Potato Chips De-oiling: After frying, use the de-oiling machine to remove the oil of the surface, and it can extent the shelf life of the product.
11. Potato Chips Flavoring: After flavored by seasoning machine, different flavor of potato chips are made. According to different tastes, there are potato chips of chicken flavor, beef flavor, spicy flavor and barbecue flavor etc.
12. Potato Chips Cooling: Cool the flavored potato chips to room temperature before packing.
13. Weighing and Packaging: The product needs quantitative packaging, net content error

Experiment 17

Development of Salted Banana Chips

Objective: *Development of Salted Banana chips.*

Banana is the second most significant natural product harvest of India close to mango. It is accessible round the year and due as it would prefer, nutritive and the rape tic worth. It is most loved product of all classes of individuals. It is a rich wellspring of sugar and nutrients especially nutrient B. Banana assists with diminishing the danger of heart illnesses. Banana is a famous organic product because of its low cost and high nutritive worth. Banana contains about 20% sugar and sensible measure of Vitamins A, B and C. This is considered to be a rich source producing food. It is consumed in several forms and among many worth added arrangements, Banana Chips is viewed as the main thing. Seared banana chips are typically delivered from under-ready banana cuts southern style in sunflower oil or coconut oil. These chips are dry (like potato chips), contain about 4% water (table), and can be salted, spiced, glossed over or jaggery covered. Some of the time banana enhancing is added.

Raw Material

1. Raw Bananas = 1.0 kg
2. Cooking Oil = 3 litre
3. Salt = 10 g
4. Water = 480 ml
5. Spices
6. Packing Material

Utensils/Materials

- Knife (stainless steel)
- Absorbent paper
- Chopping board
- Measuring cup
- Banana slicer
- Strainer
- PE/PP bags

- Colander
- Plastic sealer
- Basin/mixing bowl
- Thermometer (100°-250°C) stove
- Weighing scale

Banana chips Process Chart

Experiment 18

Production of Chocolate

Objective: *Production of Chocolate.*

The major ingredient of chocolate is cocoa mass, which is mixed with sugar, and in case of milk chocolate, also with milk powder. The conching process then follows, which is important for the development of full chocolate flavour. Melted cocoa butter is added, the chocolate mass becomes liquid and is cooled, tempered and poured into moulds to form chocolate products.

Mixing of ingredients

Sugar, cocoa mass and milk powder are intensively mixed forming a dry powder. This powder is pre-ground in various types of mills and then finely ground in a 5-roll refiner. The size of particles in chocolate is between 15-70 mm (average 25 mm). Small operators use vertical ball mills.

Optimum flow properties are required for further processing. The yield value required differs according to the potential use (e.g. hollow moulding or enrobing). The particle size affects both the yield value and the plastic viscosity. In order to reduce recipe costs, the most expensive major constituent (i.e. fat) should be used in the smallest possible proportion which will give correct flow characteristics.

Refining

Refining is essentially a grinding process during which the broken cocoa bean cotyledons (i.e. nibs) are further reduced in particle size. The process is high temperature (250C) for a short time because of the high pressure rollers used. The refining of sugar, cocoa butter and nibs promotes the formation of amorphous sugar which absorbs aromatic components and in turn masks burnt flavours from the sugar due to the high heat of refining.

The roasting process reactions may continue during refining, but it does not affect the flavour to the same degree as roasting. Steel roll refiners are used for precision grinding. An internal water circulation system ensures that the heat generated by the grinding action of the rolls is dissipated evenly and a constant temperature is maintained throughout the length of the roll.

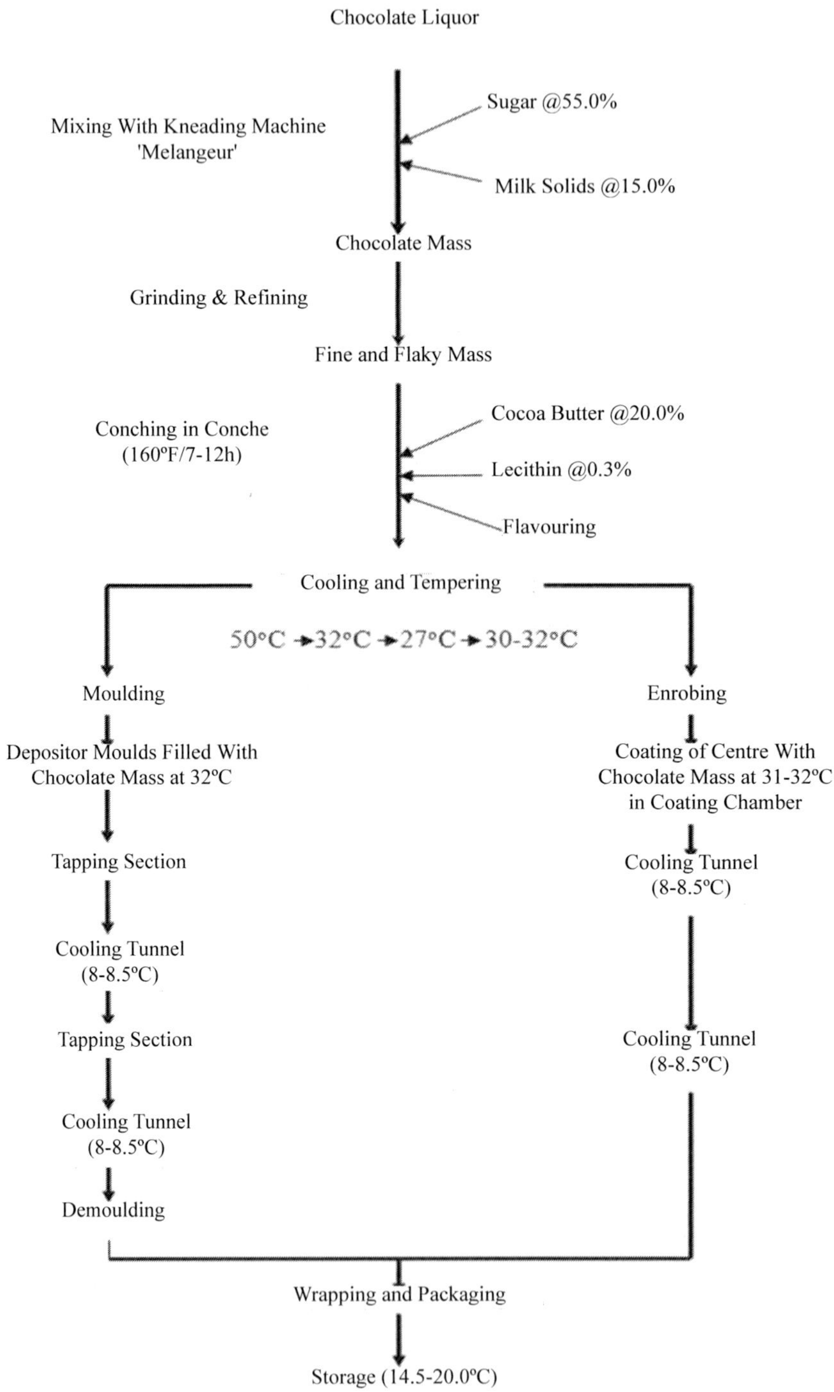

Chocolate Liquor
Mixing With Kneading Machine 'Melangeur'
Sugar @55.0%
Milk Solids @15.0%
Chocolate Mass
Grinding & Refining
Fine and Flaky Mass
Conching in Conche (160ºF/7-12h)
Cocoa Butter @20.0%
Lecithin @0.3%
Flavouring
Cooling and Tempering
50°C → 32°C → 27°C → 30-32°C
Moulding
Depositor Moulds Filled With Chocolate Mass at 32ºC
Tapping Section
Cooling Tunnel (8-8.5ºC)
Tapping Section
Cooling Tunnel (8-8.5ºC)
Demoulding
Enrobing
Coating of Centre With Chocolate Mass at 31-32ºC in Coating Chamber
Cooling Tunnel (8-8.5ºC)
Cooling Tunnel (8-8.5ºC)
Wrapping and Packaging
Storage (14.5-20.0ºC)

Uses of Chocolate

Chocolate can be consumed as such:

- Much used as an ingredient on or in other foodstuffs e.g. enrobing of biscuits. Chocolate drops go into biscuits. Chocolate used in between wafers.
- Chocolate is a popular flavour for desserts, ice cream, cakes, etc.
- In UK, milk chocolate with typical crumb caramel note is predominant.
- Cocoa flavour combines very well with many others; the popular combinations are Vanilla, peppermint, coffee and orange. Fruit flavours include nuts, cardamom, etc.

Experiment 19

Business Module of Development of Tomato Ketchup (200 kg/day) Business Model/Economic Analysis

Objective: *Business module of Development of Tomato Ketchup (200 kg/day) BUSINESS MODEL / ECONOMIC ANALYSIS.*

Fixed cost

PARTICULARS	**QUANTITY**	**RATE**	**AMOUNT (Rs.)**
Land	400m^2	Rs. 4,000/m^2	16,00,000
Building (100m^2 open &300m^2 covered area)	300m^2	Rs. 6,000/m^2	18,00,000
MACHINERIES			
Fruit Washer	1	1,50,000	1,50,000
Pulper	2	60,000*2	1,20,000
Steam Jacketed Kettle & Boiler	1	1,20,000	1,20,000
Pasteurizer	1	2,20,000	2,20,000
Filling Machine	1	50,000	50,000
Corking Machine	1	5,000	5,000
Glass Bottles	60,000	3Rs/Bottle	1,80,000
Steel Utensils (need base)			30,000
Submersible Pump	1 (2.5" bore)	40,000	40,000
Cold Store Machines			1,00,000
Generator (50KVA)			1,20,000
Pre Operation Expanses (Licence fees, DIC approval, etc.)			50,000
Miscellaneous Assets (Plastic barrels, weighing scale, etc.)			1,00,000
Insurance@1%			46,850
		TOTAL	**47,31,850/-**

Variable Cost for One Year (350 working days)

PARTICULARS	RATE	AMOUNT (Rs.) 1Yr
Fruits (Tomato) (500kg/Day)	Rs. 20/kg	35,00,000
Sugar (25 kg per day)	Rs. 40/kg	3,50,000
Spices/Preservatives & Chemicals (Rs. 42.5 kg)	Rs. 100/kg	14,87,500
Administrative Expanses		50,000
Transportation Cost		1,00,000
Advertisement		40,000
Repair & Maintenance		50,000
Electricity & Diesel	30,000/month	3,60,000
Salaries of Workers		
1. Semi skilled workers (1)	Rs. 12,000 x 1 x 12	1,44,000
2. Helper (5)	Rs. 8,500 x 5 x 12	4,80,000
3. Salesman (2)	Rs. 10,000 x 2 x 12	2,40,000
	TOTAL	**68,01,500**

Total Cost = Fixed Cost + Variable Cost = Rs. 47,31,850/- + Rs. 68,01,500/- = Rs. 1,15,33,350/-

Depreciation on building(@5% = 90,000/-

Depreciation on machinery(@10% = 1,17,500

Total Depreciation = 2,07,500/-

INCOME

Total Ketchup per day	250 kg	250 kg x 350 day = 87500 kg/year
Uncertain ketchup loss	2.5% per kg	2188 kg
Net Ketchup produced	87500-2188	85312 kg
No. of packet	850 g per packet	100367 packet per year
Packet damaged	@ 1%	1004 per year
Total safe packets		99363 packets
Cost of packet or gross income	Rs. 136 per packet	Rs 1,35,13,368/-

Total Annual Cost = Variable Cost + Depreciation = Rs. 68,01,500/- +Rs. 2,07,500/-
= Rs. 70,09,000/-

Net Income / Year = Income - Annual Cost
= Rs. 1,35,13,368/- – Rs. 70,09,000/-
= Rs. 2,05,22,368/-

Annual instalment account
To amortized even repayment plan,

As we have taken loan amount of Rs. 60,00,000/- (approximately 52% of total annual cost) (Rs. 11533500/-)

Formula for EMI Calculation is -

$$M = P\frac{r(1+r)^n}{(1+r)^n - 1}$$

where-

P = Principal loan amount = 6000000.00
n = Loan tenure in months = 120 month
r = Monthly interest rate = 9.5%
The rate of interest (R) on your loan is calculated per moonth.
r = Annual Rate of interest/12/100 = (9.5/12)/100 = 0.008

EMI = 6000000 x 0.008 x (1+0.008)^120 / [(1+0.008)^ 120]-1
= 48000 x (1.008)^120 / [(1.008)^120]-1
= 48000 x 2.6/1.60 = 78000.00 but HDFC calculator = 77639.00

Loan Repayment

Loan is Repaid by Amortized Even Repayment Plan @9.5% Interest in 10 Years.

Year	Opening balance	EMI*12	Interest yearly	Principal yearly	Closing balance
1	60,00,000	9,31,662	5,53,830	3,77,833	56,22,167
2	56,22,167	9,31,662	5,16,331	4,15,332	52,06,835
3	52,06,835	9,31,662	4,75,110	4,56,553	47,50,283
4	47,50,283	9,31,662	4,29,798	5,01,864	42,48,419
5	42,48,419	9,31,662	3,79,989	5,51,673	36,96,745
6	36,96,745	9,31,662	3,25,237	6,06,425	30,90,320
7	30,90,320	9,31,662	2,65,051	6,66,612	24,23,709
8	24,23,709	9,31,662	1,98,891	7,32,771	16,90,937
9	16,90,937	9,31,662	1,26,166	8,05,497	8,85,441
10	8,85,441	9,31,662	46,222	8,85,441	0

Net income after repayment = Rs. 2,05,22,368/- 9,31,662 = 1,95,90,706/-

Tax payment@30% = 58,77,211/-

Net profit in hand = Rs. 1,37,13,495.00

BREAKEVEN POINT

Total sale (A)	Rs. 1,35,13,368/-
Variable Cost (B)	Rs. 68,01,500/-
Total fixed cost (C)	Rs. 47,31,850/-
Contribution (D = A-B)	Rs. 6711868/-

BREAKEVEN POINT = (C/D) x 100

= (47,31,850/6711868) x 100

= 70.00% Ans.

Calculation of B : C Ratio

B : C ratio = Net income/ Total annual cost

= 2,05,22,368 / 70,09,000 = 2.93

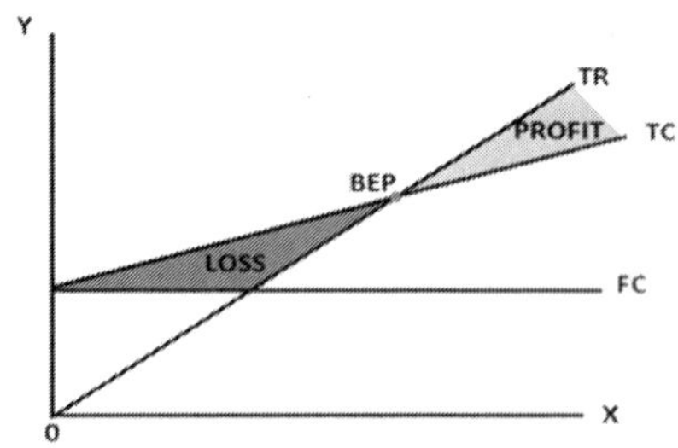

GRAPHICAL REPRESENTATION OF BEP

Experiment 20

Economic Analysis for Preparation of Banana Chips

Objective: Economic analysis for preparation of banana chips

FIXED COST

S.NO.	Particulars	Quantity	Rate(Rs)	Total Amount(Rs)
1	Land	300 m^2	8085/m^2	24,25,500
2	Construction of buildings	250 m^2	12000/m	30,00,000
3	Equipment's			
a	Peeling machine	1	34,000	34,000
b	Blancher/washer	1	1,50,000	1,50,000
c	Fryer (100Kg/h)	1	1,00,000	1,00,000
d	Sealing/packaging machine	1	20,000	20,000
4	Steel utensils (tubs, buckets etc.)		60,000	60,000
5	Miscellaneous assets			2,00,000
6	License			50,000
			Total Fixed cost	**60,39,500**

VARIABLE COST (FOR ONE YEAR)

S. No.	Particulars	Quantity	Rate(Rs)	Total amount (Rs)/year
1	Banana	400 Kg/day	1kg = 10	12,00,000
2	Frying oil	25 liter per day Total oil = 25 x 300 = 7500 lit	Rs. 130 per litre	9,75,000
3	Packaging material			2,00,000
4	Advertisement			20,000
5	Repair & maintenance			40,000
6	Transportation			1,25,000
7	salaries			
a	Skilled labour	1	12,000 Rs/month	1,44,000
b	labour	4	8,500 Rs/month	4,08,000

C	Salesman	2	10,000 Rs/month	2,40,000
d	Guard	3	8,000 Rs/month	2,80,000
8	Electricity		20,000 Rs/month	2,40,000
			Total	38,72,000

Total fixed cost = Rs. 60,395,00/-

Total variable cost = Rs. 38,72,000/-

Total fixed and variable cost = 99,11,500/-

PROJECT PROFITABILITY ANALYSIS

Recovery of banana chips = 50% of total raw materials = 20 kg chips

S. No.	Particular	Amount/year
1	Total no. of packet produced of 100 grams	600000 packets
2	Packet damage due to uncertain risk (1%)	6000 packets
	Total packets	594000 pkt
3	sales income@ Rs. 50 per, 100 grams	Rs. 2,97,00,000
4	Total income	Rs. 2,97,00,000
5	Cast of production (Variable cost)	Rs. 38,72,000
6	Depreciation on buildings@ 2% (60000) + 10% on machinery (36400)	Rs. 96,400/-
7	Instalment under amortized decreasing repayment plan	Rs. 1268647
8	Sub Total [4-(5+6+7)]	24462953
9	Tax @20%	4892590
10	Net Profit (9-10)	1,95,70,363

REPAYMENT PLAN

Repayment plan = Amortized decreasing repayment plan Loan amount = Rs 80,00,000

Interest(i)= 10%, Loan period(n)= 10 year

Year	Opening balance	EMI*12	Interest yearly	Principal yearly	Closing balance
1	8000000	1268647	777912	490735	7509265
2	7509265	1268647	726526	542121	6967144
3	6967144	1268647	669759	598888	6368256
4	6368256	1268647	607047	661600	5706656
5	5706656	1268647	537769	730878	4975779
6	4975779	1268647	461237	807410	4168368
7	4168368	1268647	376690	891957	3276412
8	3276412	1268647	283291	985356	2291056
9	2291056	1268647	180111	1088536	1202520
10	1202520	1268647	66127	1202520	0

BREAKEVEN ANALYSIS

It is the point of no loss and no profit. At this point the total cost curve and total revenue curve intersect each other.

S. No.	Particular	Amount (Rs.)
A	Total income	2,97,00,000
B	Variable cost	38,72,000
C	Contribution (A-B)	25828000
D	Fixed cost	60,39,500
E	Break-even point = D/C*100	23.38%

Experiment 21

Economic Analysis for Preparation of Multigrain Atta

Objective: Economic analysis for preparation of multigrain atta

Fixed Cost

S. No.	Particulars	Quantity	Rate (Rs)	Total (Rs)
1	Building (Rented) (*12 months)	300 msq.	25,000/month	3,00,000
2	Machinery			10,35,000
	Vibrating Screen Grain Separator	1	2,50,000	2,50,000
	Multigrain Atta Making Machine	1	1,00,000	1,00,000
	Multigrain Atta Packaging Machine	1	1,50,000	1,50,000
	Unloading Bin	1	25,000	25,000
	Storage Tank	1	10,000	10,000
	Silos (Combined)	5	5,00,000	5,00,000
3	Miscellaneous			50,000
4	License & Approval			45,000
5	Advertisement			1,00,000
			Total	**15,30,000**

- **Rent/Month** = Rs 25,000 x 12 months = Rs 3,00,000
- **Remaining Fixed Cost** = Rs 12,30,000
- **Per month Fixed Cost Used (for 5 years)** = 12,30,000/60 months
  + Rs 25,000
  = Rs 45,000

Calculation for 500 kg multigrain atta

Particular	Composition %	Quantity (kg)	Rate (□/kg)	Cost (□)
Wheat	80%	400	28	11200
Bajra	2%	10	26	260
Jowar	4%	20	30	600
Kagni	4%	20	40	800
Ragi	2%	10	36	360
Kodo	2%	10	30	300

Rajma	2%	10	140	1400
Chickpea	4%	20	75	1500
Total Cost		500		**16420**

Variable Cost (for 12 months = 300 days)

S. No.	Particulars	Rate (Rs)/month	Annual cost (Rs.)
1	Raw Material Consumed Capacity: 500 kg per day	16420* 300	49,26,000
2	Electricity Consumed	25,000	3,00,000
3	Wages & Labour		
3a.	Plant Operator	20,000	2,40,000
3b.	Supervisor	18,000	2,16,000
3c.	Skilled Labour (2)	12,000	2,88,000
3d.	Unskilled Labour (2)	10,000	2,40,000
3e.	Manager	20,000	2,40,000
3f.	Administrative Staff (1)	12,000	1,44,000
3g.	Accountant (1)	12,000	1,44,000
4	Repair Maintenance & Packaging	30,000	3,60,000
5	Depreciation @ 2% on machine	20,700	2,48,000
		Total	**73,46,400**

Project Profitability Analysis (MONTHLY) / Project Profitability Analysis (YEARLY)

S. No.	Particular	Total
1	Total no. of packets produced (500g per packet) 500*2*300 = 300000 pkt	300000 pkt
2	Total packets damaged @ 1%	3000 pkt
	Packet of sale (1-2)	297000 pkt
3	Sales income @ Rs. 55 per 500g packet	Rs. 1,63,35,000
4	Total income	Rs. 1,63,35,000
5	Cost of production (variable cost)	Rs. 73,46,400
6	Fixed cost	Rs. 15,30,000
7	Loan repayment @ 10%	Rs. 12,68,647
8	Profit [4-(5+6+7)]	Rs. 61,89,953
9	Tax @ 20%	Rs. 1237990
10	Net profit (in hand)	Rs. 4951953

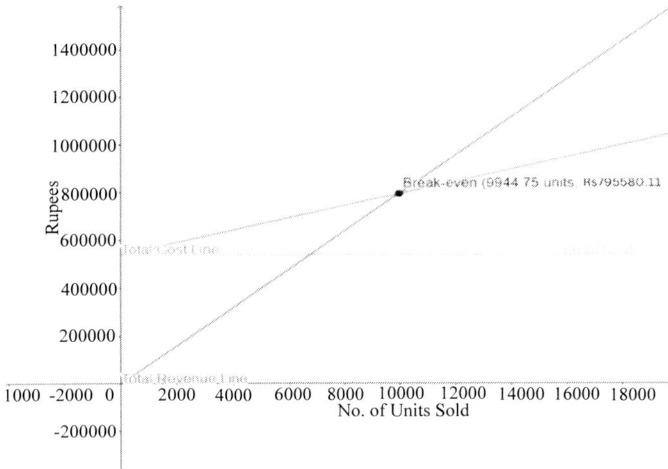

Since the break-even point is **less than** the units sold (U), the company **makes a profit**.

REPAYEMENT PLAN

Repayment plan = Amortized decreasing repayment plan Loan amount = Rs. 80,00,000

Interest(i) = 10%, Loan period(n) = 10 year

Year	Opening balance	EMI*12	Interest yearly	Principal yearly	Closing balance
1	8000000	1268647	777912	490735	7509265
2	7509265	1268647	726526	542121	6967144
3	6967144	1268647	669759	598888	6368256
4	6368256	1268647	607047	661600	5706656
5	5706656	1268647	537769	730878	4975779
6	4975779	1268647	461237	807410	4168368
7	4168368	1268647	376690	891957	3276412
8	3276412	1268647	283291	985356	2291056
9	2291056	1268647	180111	1088536	1202520
10	1202520	1268647	66127	1202520	0

Break-Even Point Analysis

The Break-Even Point represents the quantity of units that need to be sold in order to cover all costs (both fixed and variable), resulting in neither profit nor loss.

S. No.	Particular	Amount (Rs.)
A	Total income	1,63,35,000
B	Variable cost	73,46,400
C	Contribution (A-B)	8988600
D	Fixed cost	15,30,000
E	Break-even point = D/C*100	17.02%

$$\text{Break even point} = \frac{\text{fixed cost}}{\text{contribution}} \times 100 = \frac{5,40,000}{3,352,2000} \times 100 = 16.1\%$$

It needs to sell **4212 pkt per month or 170 pkt per day** to break even.

References

Kalia, M (2004). Food Analysis and Quality Control. Kalyani Publishers, Ludhiana.

Kalia, M and Sood S (2002). Food Preservation and Processing. Kalyani Publishers, Ludhiana.

Manay NS and Shadaksharaswany M (2002). Foods: Facts and Principles. New Age International (P) Ltd Publisher, New Delhi

Mohsenin NN (1996). Physical Properties of Plant and Animal materials. Gordon and Breach Publisher. Australia

Nagi HPS, Sharma Savita and Sekhon KS (2007). Handbook of Cereal Technology: A Practical Approach. Kalyani Publishers, Ludhiana.

Ranganna S (2001). Handbook of Analaysis and Quality Control of Fruit and Vegetable Products. Tata McGraw Hill Pub. Co. Ltd. New Delhi

Sharma SK, Steven J. Mulvaney and Syed SH Rizvi (2003). Food Process Engineering: Theory and Laboratory Experiments. A John Wiley & Sons Inc. New York.

Srivastava RP and Kumar, Sanjeev (1998). Fruit and Vegetable Preservation: Principle and Practice. International Book Distributing Co. Lucknow

Notes

Index